U0053387

Sky High

A Photographic History of
the Flying Tigers and the Black Cat Squadron

圖說飛虎隊與黑貓中隊

主編 / 楊善堯

翻譯 / 廖彥博

作者 / Bennett Quo、Everett Wang、Nathan Yao、Patrick Hao

誌閎人文

目次

主編序
Editor's Words

This book is an amalgamation of military history, archival images, historical studies and humanities education, and international relations between the Republic of China and the United States, all rolled into a single photograph album.

In June 2023, a group of Chinese-American high school students and their parents traveled to Taiwan under the leadership of Dr. Hsiao-ting Lin, Research Fellow and Curator of the Modern China and Taiwan Collection for the Hoover Institution at Stanford University. Together, the group visited the Presidential Palace, Academia Historica, National Revolutionary Martyrs' Shrine, Chiang Kai-shek Memorial Hall, Palace Museum, Grand Hotel, Fu Jen Catholic University, Longshan Temple, Bopiliao Historic Block, Daxi Old Street, and other attractions with strong ties to "Republican, Chinese, and local Taiwanese culture." Experiencing these sites first hand not only allowed the students to develop an understanding of their historical and cultural significance, but also resonated with them on a personal level; these are all places they had learned about while growing up in the United States and are intimately connected to their own family history. Understandably, this tour resulted in considerable culture shock for this group of high school students visiting Taiwan(Republic of China) for the first time. During this visit, they consistently concluded that their cultural homeland, the Republic of China, is tightly intertwined with the country in which they grew up, the United States. This realization motivated them and sparked their interest in further exploring this period of history, which, in turn, was the inspiration for this book.

The military cooperation between the Republic of China and the United States in the early-twentieth century can be traced back to the later stages of the Second Sino-Japanese War (1937-1945). As the Chinese government waged a full-scale and arduous war against Japan, the United States gave the Republic of China considerable support, from civilian assistance to formal military aid. In the 1950s, again with U.S. support, Taiwan's economy recovered and developed rapidly, while American military cooperation with the Republic of China was an especially vital part of Cold War history. Moreover, the two main subjects of this book, the Flying Tigers and the Black Cat Squadron, connect these historical themes together and offer a significant illustration of Sino-American military cooperation.

The four authors of this book, Bennett Quo, Everett Wang, Nathan Yao, and

Patrick Hao, employ chronologically arranged historical photographs and archival materials from the Republic of China's Academia Historica, the U.S. National Archives, and the Zhe Hong Humanities Studio collections to provide a clear and vibrant portrait of these two subjects. In so doing, the visual impact of these images evokes a much greater response in the reader than that of words alone. It is hoped that the publication of this photograph album, as well as highlighting the process that four high school students undertook to interpret materials from historical archives, will attract more attention to this period of history, and simultaneously enable readers to understand that historical archives are not only for scholars, but for everyone. . Finally, this book is intended to demonstrate the practical application of historical studies and the humanities to encourage even more development in these fields.

Yang, Shan-Yao
2023.10.10

這本書是一本結合軍事歷史、檔案影像、歷史人文教學、中華民國與美國國際關係的攝影冊。

　　2023 年 6 月，一群美籍華裔高中生與家長，在史丹佛大學胡佛研究所林孝庭研究員的帶領下來到了臺灣，在總統府、國史館、忠烈祠、中正紀念堂、故宮博物院、圓山飯店、輔仁大學、龍山寺、剝皮寮、大溪老街等地方進行了參訪與交流，這些帶有濃厚「民國風、中華文化、臺灣在地文化」風格的景點，以及這些景點帶來的實體視覺與景點背後所隱含的歷史文化意義，都與他們從小在美國所學以及家庭文化有所關聯，給了這群第一次來到臺灣（中華民國）的高中生們相當大的文化衝擊。在參訪的過程中，他們不斷地發現到一個問題：「中華民國跟他們所生長的美國有著高度關聯性」，這引起了他們去探索這段歷史過往的動機與興趣，也是這本書誕生的緣起。

　　有關中華民國與美國在二十世紀的軍事往來，可從抗日戰爭的中後期開始談起。在中華民國政府進行全面性的艱苦戰爭時，美國從民間援助到正式軍援，給了當時政府相當大的支持，到了 1950 年代後，在美援支持下臺灣快速的全面復甦，尤其在軍事上的援助，更是冷戰時代下中華民國與美國的重要連結。本書的飛虎隊與黑貓中隊這兩個主題，正好是串聯起這段過往的重要歷史。

　　Bennett Quo、Everett Wang、Nathan Yao、Patrick Hao 四位作者利用國史館、美國國家檔案館與喆閎人文工作室所典藏的歷史照片與檔案，透過照片解讀與時序排列，清楚且豐富的呈現出這兩個主題。視覺影像的渲染力與感受力遠大於單純文字給讀者帶來的感受，希望這本攝影冊的面世，透過四位高中生解讀歷史檔案的過程，能引起更多人對於這段歷史的注意，以及讓歷史人文融入應用實作有更多的成果展現。

楊善堯

民國一一二年十月十日

A Brief History of the
Flying Tigers

The Flying Tigers, or the American Volunteer Group (AVG) was a legendary group of aviators, holding a remarkable, yet often overlooked place in military history. Emerging during the tumultuous times of the late 1930s, their contributions not only altered the course of World War II but also left an indelible mark on the art of aerial combat. This book, through its unique collection of photographs, will explore the origins, operations, and legacy of the Flying Tigers.

With both the Sino-Japanese conflict in full swing and the outbreak of the Second World War in Europe, the Republic of China found itself in dire need of assistance. Drawing upon experienced American aviators, General Claire Lee Chennault sought to help defend China's airspace against the skilled Japanese pilots. By offering high salaries and bonus rewards for each enemy kill, he was able to raise a formidable aircrew before his debut in China.

In the spring of 1941, General Claire Lee Chennault arrived in China with the American volunteer pilots to aid the Chinese Air Force. While the volunteers were mostly former U.S. military aviators, a small percentage were commercial civilian pilots.

After gathering in Burma for rigorous training, the Flying Tigers faced numerous challenges before engaging in combat. However, delayed by logistical complexities and the shifting international policies, the squadron's first combat didn't occur until December 20, 1941, 13 days after the infamous attack on Pearl Harbor.

Adorned with the distinctive Chinese colors of red and gold, the brand-new Curtiss P-40 Warhawk became the mainstay of the Flying Tiger's fleet. Though heavily outnumbered, and undersupplied, the Flying Tigers swiftly gained recognition for their exceptional combat performance. Despite the constant defeats by the Japanese that dominated news headlines, the AVG aerial victories enabled the Chinese to stabilize the front and slow down the Japanese invasion. Throughout the duration of the war, the aviators of the Flying Tigers managed to shoot down 296 enemy aircraft while only losing 14 pilots themselves.

Though the Flying Tigers' played a key role in securing the airspace over China, their legacy extends far beyond their combat achievements. Their arrival in China, in April of 1941 lifted spirits and demonstrated that victory against seemingly insurmountable odds was attainable. One such example of the missions undertaken by the Flying Tigers was the "morale flights," in which members of the Flying Tigers would fly over entrenched Chinese troops, offering air support. The heroism that the 1st AVG displayed not only inspired many in the United States but also significantly boosted the morale of the beleaguered Chinese soldiers. The squadron's impact was so profound that it led to a renewed focus on countering Japan's air superiority, paving the way for America's own domination of skies in the closing days of the war.

One instance of the Flying Tiger's impact was their key role in supplying the Chinese during World War II. At the eastern end of the Himalayas, the Hump was vital to the fight in Asia. It was a vital supply route that enabled the Allies to keep the Chinese forces supplied during World War II. After the Burma road was cut off by Japanese forces, the Hump became the sole route in or out of China. Flying through the Hump posed incredible risk and required tremendous resilience from its pilots. The low-lying jungles and rolling hills of Burma abruptly gave way to the Santsung range. With many of its peaks rising over 15,000 feet above sea level, flying through the hump was a daunting task. Even the most skilled pilots struggled to get the cargo-laden planes high enough to cross the Hump. Moreover, the constant turbulence and unpredictable weather conditions further contributed to the notoriety of the route. The Hump was so vital to the Allied cause that pilots were ordered to take off regardless of the weather. While the Hump claimed the lives of upwards of 1300 airmen, their bold sacrifice in resupplying China undoubtedly saved millions more.

With their mission fulfilled, the Flying Tigers were officially disbanded on July 4, 1942. However, their spirit endured. Under General Chennault's leadership, the 23rd Fighter Group of the United States Army Air Forces continued to achieve similar combat success, carrying forward the iconic P-40 nose art. The Flying Tigers' ethos of valor, sacrifice, and unity remained a source of inspiration for the many aviators who succeeded them.

While the contributions of the Flying Tigers are often overshadowed by the larger events that surrounded them, they represent the courage and determination of America to aid its allies. Assembled from different branches of the U.S. military and propelled by a shared purpose, these American aviators contributed far above what was expected of them. Standing as a testament to the fighting spirit of the Allied Powers, their legacy continues to remind us that heroes can emerge from unexpected places, shaping the course of history and protecting freedom and liberty.

飛虎隊

被稱為「飛虎隊」的中華民國空軍美籍志願大隊（American Volunteer Group）是一支傳奇的飛行部隊，在軍事史上具有重要的地位，卻經常受到忽略。這支部隊成軍於一九三〇年代後期的動盪時代，他們的貢獻不僅改變了第二次世界大戰的進程，更在空戰技術方面留下了不可抹滅的印記。本書將透過珍貴的歷史照片集，探討飛虎隊的成軍起源、作戰任務，以及其遺澤影響。

隨著對日抗戰全面展開，第二次世界大戰在歐洲爆發，中華民國急需外援。此時陳納德（Claire Lee Chennault）將軍招募富有經驗的美國飛行員，協助中國捍衛領空，以抵禦技術熟練的日本飛行員。陳納德以提供高薪和每擊落一架敵機便有額外賞金做號召，很快就組建起一支實力堅強的機組人員隊伍。

一九四一年春季，陳納德將軍偕同美籍志願空軍人員一同抵達中國，援助中華民國空軍。大多數志願大隊的隊員都是退役的美軍飛行員，不過也有少部分人是商用飛機的駕駛。

飛虎隊先在緬甸集結接受訓練，但是還未投入作戰，便遭遇重重挑戰。由於後勤補給的複雜與國際局勢的變化，飛虎隊一直到一九四一年十二月二十日（也就是在惡名昭彰的日軍偷襲珍珠港十三天之後）才第一次投入戰鬥。

全新的寇蒂斯 P-40「戰鷹」戰鬥機（Curtiss P-40 Warhawk）漆以中國傳統的紅、金兩色，成為飛虎隊的主力機種。儘管飛機數量嚴重不足，地勤補給也極度缺乏，飛虎隊卻很快因其出色的戰鬥表現而贏得敬重。雖然日軍攻城掠地的消息持續佔據著新聞頭條，但飛虎隊在空中取得的勝利使中國穩住了戰線，並拖延了日本侵略的腳步。在整個作戰期間，中國空軍美籍志願大隊以自身僅損失十四名飛行員的代價，成功擊落了二百九十六架敵機。

誠然飛虎隊在保衛中國領空方面發揮了關鍵作用，但他們為後世帶來的影響遠遠大過於作戰所取得的勝果。一九四一年四月他們抵達中國，便使中國軍民士氣為之一振，而且證明了看似堅不可催的強敵是可以戰勝的。其中一個這樣的例子，是飛虎隊需要執行「鼓舞士氣」的飛

行任務，也就是飛越堅守陣地的中國軍隊上空，提供空中支援。美籍志願航空隊展現出的英雄氣概不僅鼓舞了許多美國人，也大大提振了陷入困境的中國軍隊士氣。飛虎隊造成的影響極其深遠，不僅使中國當局重新注意與日本空中優勢相抗衡，更為美國在戰爭後期完全主導制空權奠定基礎。

飛虎隊在第二次世界大戰中為中國提供補給起到的關鍵作用，是他們造成深遠影響的又一個例證。所謂「駝峰」（Hump）航線位於喜馬拉雅山東側，對於亞洲戰場至關緊要。這條重要的航線，使得盟軍能為中國軍隊提供物資補給。在滇緬公路被日軍切斷以後，駝峰航線成為進出中國的唯一途徑。飛越駝峰航線面臨著極大的風險，考驗飛行機組人員強大的應變能力。原本緬甸低海拔的熱帶叢林與平緩的山丘，一下子變成崇山峻嶺。由於許多山峰海拔高度超過一萬五千英呎，飛越「駝峰」航線這些崇山峻嶺是一項極其艱鉅的任務。即便是技術最純熟的飛行員，也不能保證滿載貨物的飛機高度能越過這些層峰疊嶂的山峰。此外，持續出現的亂流和難以預測的天氣狀況，使得這條航線更是聲名狼藉。但是「駝峰」航線對盟軍的戰局至關重要，因此飛行員常接到命令，無論天氣狀況如何都必須起飛。雖然「駝峰」航線奪走了超過一千三百位機組人員的生命，但他們為了運補中國而做出的英勇犧牲，無疑拯救了數以百萬計的生靈。

隨著使命的完成，飛虎隊於一九四二年七月二日正式解散。然而，他們的精神永續長存。在陳納德將軍的領導下，承接在華作戰任務的美國陸軍航空隊第二十三大隊（the 23rd Fighter Group of the United States Army Air Forces）繼續獲致與飛虎隊相似的戰果，並將 P-40 機頭彩繪持續發揚光大。飛虎隊勇敢、犧牲與團結的精神，仍然激勵著眾多後繼者追隨效法。

儘管飛虎隊的功績經常被同時期更重大的歷史事件掩蓋，但是飛虎隊畢竟代表了美國援助盟友的勇氣與決心。這些美國飛行員來自美軍的不同單位，為了共同一致的目標而努力，所做出的貢獻遠超出人們所預期。作為盟軍戰鬥精神的見證，他們留下的遺澤不斷地提醒我們：英雄可以從意想不到的地方出現，塑造歷史進程，並且護衛自由和解放。

Flying Tigers.
飛虎隊員合影

1942
《陳誠副總統文物》 *Vice-President Chen Cheng's Collection*
典藏號 Archive No.：008-030400-00002-013

圖片：國史館 photo courtesy of Academia Historica

Commander Claire Chennault appreciated Flying Tigers pilots for their hard work.

慰勞飛虎隊

1942
《陳誠副總統文物》 *Vice-President Chen Cheng's Collection*
典藏號 Archive No.：008-030400-00002-013

圖片：國史館 photo courtesy of Academia Historica

Major General Claire Chennault, the commander
of the Fourteenth Air Force of the United States.
陳納德將軍

1942
《陳誠副總統文物》 *Vice-President Chen Cheng's Collection*
典藏號 Archive No.：008-030400-00002-013

圖片：國史館 photo courtesy of Academia Historica

A P-40 aircraft was being pushed onto the
runway to prepare for takeoff.

一架 P-40 飛機正推上跑道準備起飛

1942
典藏號 Archive No.：jdsyxzl-zdp-001050

圖片：美國國家檔案館 National Archives

Brigadier General Chennault was conducting a combat mission briefing with members of the Flying Tigers.

陳納德准將正在跟飛虎隊隊員進行作戰任務說明

1942
典藏號 Archive No.：jdsyxzl-zdp-001090

圖片：美國國家檔案館 National Archives

American ground crews were performing
maintenance works.

美軍地勤人員正在進行維修作業

1942
典藏號 Archive No.：jdsyxzl-zdp-001050

圖片：美國國家檔案館 National Archives

The Flying Tigers planes the Curtiss P-40 soaring through the skies.

飛虎隊的座機、寇帝斯公司生產的 P-40
戰鬥機翱翔天際

1942
典藏號 Archive No.：jdsyxzl-zdp-001024,
　　　　　　　　　　jdsyxzl-zdp-001026

圖片：美國國家檔案館 National Archives

Chinese people paid tribute to the Flying Tigers pilots.
慰勞飛虎隊

1942
《陳誠副總統文物》 *Vice-President Chen Cheng's Collection*
典藏號 Archive No.：008-030400-00002-013

圖片：國史館 photo courtesy of Academia Historica

Chinese people paid tribute to the Flying Tigers pilots.
慰勞飛虎隊

1942
《陳誠副總統文物》 *Vice-President Chen Cheng's Collection*
典藏號 Archive No.：008-030400-00002-013

圖片：國史館 photo courtesy of Academia Historica

General Chennault in a photo with General
Chen Cheng and other Chinese and American
generals in Kunming.

陳納德在昆明與陳誠等中美將領合影

1943
《陳誠副總統文物》 *Vice-President Chen Cheng's Collection*
典藏號 Archive No.：008-030400-00003-010

圖片：國史館 photo courtesy of Academia Historica

General Chennault in a photo with General
Chen Cheng and other Chinese and American
generals in Kunming.

陳納德在昆明與陳誠等中美將領合影

1943
《陳誠副總統文物》 *Vice-President Chen Cheng's Collection*
典藏號 Archive No.：008-030400-00003-010

圖片：國史館 photo courtesy of Academia Historica

Chen Cheng accompanied Bai Chongxi to visit the U.S. Army's Fourteenth Air Force training.

陳誠陪同白崇禧參觀美國陸軍第十四航空隊訓練

1943/08/11
《陳誠副總統文物》 *Vice-President Chen Cheng's Collection*
典藏號 Archive No.：008-030400-00008-004

圖片：國史館 photo courtesy of Academia Historica

Chen Cheng accompanied Bai Chongxi to visit
the U.S. Army's Fourteenth Air Force training.
陳誠陪同白崇禧參觀美國陸軍第十四航空
隊訓練

1943/08/11
《陳誠副總統文物》 *Vice-President Chen Cheng's Collection*
典藏號 Archive No. ：008-030400-00008-004

圖片：國史館 photo courtesy of Academia Historica

Chen Cheng accompanied Bai Chongxi to visit the U.S. Army's Fourteenth Air Force training.

陳誠陪同白崇禧參觀美國陸軍第十四航空隊訓練

1943/08/11

《陳誠副總統文物》 *Vice-President Chen Cheng's Collection*

典藏號 Archive No.：008-030400-00008-004

圖片：國史館 photo courtesy of Academia Historica

Chen Cheng accompanied Bai Chongxi to visit
the U.S. Army's Fourteenth Air Force training.

陳誠陪同白崇禧參觀美國陸軍第十四航空
隊訓練

1943/08/11
《陳誠副總統文物》 *Vice-President Chen Cheng's Collection*
典藏號 Archive No.：008-030400-00008-004

圖片：國史館 photo courtesy of Academia Historica

Chen Cheng accompanied Bai Chongxi to visit
the U.S. Army's Fourteenth Air Force training.
陳誠陪同白崇禧參觀美國陸軍第十四航空
隊訓練

1943/08/11
《陳誠副總統文物》 *Vice-President Chen Cheng's Collection*
典藏號 Archive No.：008-030400-00008-004

圖片：國史館 photo courtesy of Academia Historica

Chen Cheng accompanied Bai Chongxi to visit the U.S. Army's Fourteenth Air Force training.

陳誠陪同白崇禧參觀美國陸軍第十四航空
隊訓練

1943/08/11
《陳誠副總統文物》 *Vice-President Chen Cheng's Collection*
典藏號 Archive No.：008-030400-00008-004

圖片：國史館 photo courtesy of Academia Historica

Chen Cheng accompanied Bai Chongxi to visit
the U.S. Army's Fourteenth Air Force training.

陳誠陪同白崇禧參觀美國陸軍第十四航空
隊訓練

1943/08/11
《陳誠副總統文物》 *Vice-President Chen Cheng's Collection*
典藏號 Archive No. ：008-030400-00008-004

圖片：國史館 photo courtesy of Academia Historica

Chen Cheng accompanied Bai Chongxi to visit
the U.S. Army's Fourteenth Air Force training.

陳誠陪同白崇禧參觀美國陸軍第十四航空
隊訓練

1943/08/11
《陳誠副總統文物》 *Vice-President Chen Cheng's Collection*
典藏號 Archive No.：008-030400-00008-004

圖片：國史館 photo courtesy of Academia Historica

Chen Cheng accompanied Bai Chongxi to visit
the U.S. Army's Fourteenth Air Force training.

陳誠陪同白崇禧參觀美國陸軍第十四航空
隊訓練

1943/08/11
《陳誠副總統文物》 *Vice-President Chen Cheng's Collection*
典藏號 Archive No.：008-030400-00008-004

圖片：國史館 photo courtesy of Academia Historica

Chairman of the National Government Chiang
Kai-shek and his wife took a photo with General
Chennault, Commander of the Fourteenth Air
Force of the United States.
國民政府主席蔣中正伉儷與美國第十四航
空隊司令官陳納德合影

1945/03/24
《蔣中正總統文物》 *President Chiang Kai-shek's Collection*
典藏號 Archive No.：002-050101-00005-032

圖片：國史館 photo courtesy of Academia Historica

Chairman of the National Government Chiang
Kai-shek chaired a discussion with the U.S.
Fourteenth Air Force in March 24, 1945.

國民政府主席蔣中正主持美國第十四航空
隊座談

1945/03/24
《蔣中正總統文物》 *President Chiang Kai-shek's Collection*
典藏號 Archive No.：002-050101-00005-036

圖片：國史館 photo courtesy of Academia Historica

Chairman of National Government Chiang Kai-shek took a group photo with senior Chinese and American generals.

國民政府主席蔣中正與中美高級將領合影

1945/03/24
《蔣中正總統文物》 *President Chiang Kai-shek's Collection*
典藏號 Archive No.：002-050101-00005-040

圖片：國史館 photo courtesy of Academia Historica

Chairman of the National Government Chiang
Kai-shek and General Chennault, Commander
of the Fourteenth Air Force of the United States.

國民政府主席蔣中正與美國第十四航空隊
司令官陳納德准將合影

1945/03/24
《蔣中正總統文物》 *President Chiang Kai-shek's Collection*
典藏號 Archive No.：002-050101-00005-028

圖片：國史館 photo courtesy of Academia Historica

Accompanied with Major General Claire Chennault, Chairman of the National Government Chiang Kai-shek inspected the Fourteenth Air Force of the U. S. Army, and introduced the officers of the Fourteenth Air Force one by one.

國民政府主席蔣中正在陳納德准將陪同下視察美國第十四航空隊，並逐一引介各位軍官

1945/03/24
《蔣中正總統文物》 *President Chiang Kai-shek's Collection*
典藏號 Archive No.：002-050101-00005-026

圖片：國史館 photo courtesy of Academia Historica

Observers attending the medal ceremony
for General Chennault, Commander of the
Fourteenth Air Force of the United States.
參加美國第十四航空隊隊長陳納德將軍受
勳儀式之觀禮人員

1945/07/30
《蔣中正總統文物》 *President Chiang Kai-shek's Collection*
典藏號 Archive No. ：002-050101-00005-115

圖片：國史館 photo courtesy of Academia Historica

Chiang Kai-shek, Chairman of the National Government, presented the Order of the Blue Sky and White Sun to General Chennault, commander of the Fourteenth Air Force of the United States, in recognition of his contribution in assisting the Anti-Japanese War and took a photo of congratulations.

國民政府主席蔣中正向美國第十四航空隊隊長陳納德將軍贈授青天白日勳章以酬其協助對日抗戰之功蹟後握手道賀留影

1945/07/30

《蔣中正總統文物》 *President Chiang Kai-shek's Collection*

典藏號 Archive No.：002-050101-00005-114

圖片：國史館 photo courtesy of Academia Historica

President Chiang Kai-shek and Madame Chiang
took a photo with the American General Claire
Chennault at the Kaigetang (Victory Chapel) on
Zhongshan mountain in Nanjing.

總統蔣中正伉儷與美國陳納德將軍於南京
小紅山凱歌堂亭留影

1948/08/01
《蔣中正總統文物》 *President Chiang Kai-shek's Collection*
典藏號 Archive No.：002-050113-00004-141

圖片：國史館 photo courtesy of Academia Historica

In 1949, Chiang Kai-shek, who had stepped down from the presidency but still hold the position as the Director-General of the Chinese Nationalist Party (Kuomintang), talked with Claire L. Chennault, the captain of the American Civil Aviation Corps.

中國國民黨總裁蔣中正與美國民航隊隊長陳納德談話留影

1949/10/17
《蔣中正總統文物》 *President Chiang Kai-shek's Collection*
典藏號 Archive No.：002-050101-00012-135

圖片：國史館 photo courtesy of Academia Historica

Song Meiling (Madame Chiang) took photo with General and Mrs. Chennault and their two daughters.

宋美齡與民航空運公司陳納德夫婦合影

1956/03/19
《蔣中正總統文物》*President Chiang Kai-shek's Collection*
典藏號 Archive No.：002-050113-00009-273

圖片：國史館 photo courtesy of Academia Historica

Song Meiling (Madame Chiang Kai-shek) receives Mrs. Anna Chennault (Chen Xiangmei), wife of the Chairman of Civil Air Transport Claire Chennault.

宋美齡接待民航空運公司陳納德夫人陳香梅

1956/03/19

《蔣中正總統文物》*President Chiang Kai-shek's Collection*

典藏號 Archive No.：002-050113-00009-269

圖片：國史館 photo courtesy of Academia Historica

陳納德將軍銅像揭幕
蔣夫人昨主持典禮
艾森豪、尼克森等均寄頌詞

（本報訊）故虎將軍陳納德上將紀念銅像，昨天上午十時三十分由蔣總統夫人主持介壽路新公園重舉行。

銅像揭幕典禮，故虎將軍陳納德上將在一九五八年七月二十七日因癌病逝世於美國。陳納德將軍生前曾參與美國志願隊援助我國對日作戰，於一九四二年因病情加重離開中國，在紐約渡過了他餘年，一九四七年任在上海重組航空運輸公司，他於一九四七年十二月二十一日和陳香梅女士在上海結婚任一時政府召開之軍事會議。

1960/04/14
《蔣中正總統文物》President Chiang Kai-shek's Collection
典藏號 Archive No.：002-110101-00013-014

圖片：國史館 photo courtesy of Academia Historica

Soong Meiling and Anna Chennault attended
the unveiling ceremony of the bronze statue of
General Chennault in the Taipei New Park.

宋美齡偕陳香梅參加新公園陳納德將軍銅
像揭幕式

1960/04/14
《蔣中正總統文物》 *President Chiang Kai-shek's Collection*
典藏號 Archive No.：002-050113-00014-167

圖片：國史館 photo courtesy of Academia Historica

Guests at the unveiling ceremony of the bronze
statue of General Chennault all rose to welcome
Song Meiling.

陳納德將軍銅像揭幕式觀禮來賓起立歡迎
宋美齡蒞臨

1960/04/14

《蔣中正總統文物》 *President Chiang Kai-shek's Collection*

典藏號 Archive No.：002-050113-00014-168

圖片：國史館 photo courtesy of Academia Historica

Photo of Song Meiling and Anna Chennault (nee
Chen Xiangmei) taking a break and talking at
the unveiling ceremony of the Bronze Statue of
General Chennault.

宋美齡於陳納德將軍銅像揭幕式中偕陳香
梅休息談話留影

1960/04/14
《蔣中正總統文物》 *President Chiang Kai-shek's Collection*
典藏號 Archive No.：002-050113-00009-269

圖片：國史館 photo courtesy of Academia Historica

A glimpse of the venue for the unveiling
ceremony of the bronze statue of General
Chennault.

陳納德將軍銅像揭幕典禮會場一瞥

1960/04/14

《蔣中正總統文物》 *President Chiang Kai-shek's Collection*

典藏號 Archive No.：002-050113-00014-173

圖片：國史館 photo courtesy of Academia Historica

Soong Meiling personally unveiled the bronze
statue of General Chennault.

宋美齡親為陳納德將軍銅像行揭幕式

1960/04/14
《蔣中正總統文物》 *President Chiang Kai-shek's Collection*
典藏號 Archive No.：002-050113-00014-174

圖片：國史館 photo courtesy of Academia Historica

Song Meiling (Madame Chiang Kai-shek) and Anna Chennault (nee Chen Xiangmei) left the site after the ceremony of unveiling the bronze statue of General Claire Chennault.

宋美齡與陳香梅於陳納德將軍銅像揭幕典禮完成後相偕離去時留影

1960/04/14
《蔣中正總統文物》 *President Chiang Kai-shek's Collection*
典藏號 Archive No.：002-050113-00014-175

圖片：國史館 photo courtesy of Academia Historica

總統事略日記

中華民國五十三年七月二日

總統是日在

大事記略

九時三十分　作戰會談

十一時　接見美國國際合眾社副社長賀伯列等二人

十一時二十分　接見菲華青年醫師回國服務考察團歐陽士端等

十八人（見附件二）

十一時三十分　接見唐君鉑

二十時　晚宴款待前中國空軍美國志願大隊隊員暨眷屬（見附件三）

言論記略

總統對合眾社記者指出美軍在東南亞地區肅清已與共匪作戰

（見附件一）

總統伉儷昨晚設宴
款待飛虎隊隊員
蔣夫人並勗勉該隊隊員
繼續與我合作奮鬥抗暴

〔中央社訊〕蔣總統暨夫人，二日晚八時，蒞臨圓山大飯店麒麟廳，與前在華被稱為「飛虎隊」的中國空軍美國志願大隊隊員眷屬，並設宴款待。

飛虎隊隊員眷屬大多數都是戰後結婚的名譽司令，即席該隊隊員致詞初次來華，接見其。飛虎隊隊員於二次世界大戰時，克敵致勝，歷建功勳，他們都曾，該隊隊員暨眷屬，都添為興奮與感謝。蔣總統納該隊隊員，雖已退休，但希望能繼續與我國攜手合作，以建鬥合作，奮鬥抗暴。

蔣夫人並勗勉該隊隊員，繼續與我合作奮鬥抗暴。

夫人致故蔣納該隊隊員，受已故陳納德將軍識拔，

（蔣夫人致詞全文刊三版）
（蔣夫人杜絕邪惡。詞意懇切，）

In the evening, the President hosted a banquet for members and family members of the "Flying Tigers," American Volunteer Group.

總統蔣中正伉儷設宴款待「飛虎隊」中國空軍美國志願大隊隊員暨眷屬

1964/07/02

《蔣中正總統文物》 *President Chiang Kai-shek's Collection*

典藏號 Archive No.：002-110101-00030-002

圖片：國史館 photo courtesy of Academia Historica

President Chiang Kai-shek and his wife, Song
Meiling, hosted a banquet for former American
Flying Tigers members.

總統蔣中正伉儷款宴美國前飛虎隊員

1964/07/02
《蔣中正總統文物》 *President Chiang Kai-shek's Collection*
典藏號 Archive No.：002-050101-00055-183

圖片：國史館 photo courtesy of Academia Historica

President Chiang Kai-shek and his wife, Song Meiling, hosted a banquet for former American Flying Tigers members.

總統蔣中正伉儷款宴美國前飛虎隊員

1964/07/02

《蔣中正總統文物》 *President Chiang Kai-shek's Collection*

典藏號 Archive No.：002-050101-00055-188

圖片：國史館 photo courtesy of Academia Historica

14TH AIR FORCE ASSOCIATION, INC.

Flying Tigers

May 27, 1968

Minister Konsin C. Shah
Department of Protocol
Ministry of Foreign Affairs
Taipei, Taiwan

Dear Minister Shah:

In compliance with your request of April 29, I submit herewith the
latest list of our members who will attend the convention in Taiwan.
There still exists the possibility of a few last minute additons and
the chance of one or two cancellations. Of these I shall keep you
informed up until the date of our departure.

By way of explanation, the symbol (M) prior to a name indicates that
the person is a member of the 14th AFA. Those not so designated are
wives and children. The symbol (T) after a name indicates that he
will be with us in Taiwan only and will not be making the entire 21-
day tour with us. (TA) indicates a representative of the travel
agency which has arranged our trip. I have put an asterisk after
the name of each of our officers. Please refer below. I wish also
to point out that this list does not include our members who reside
in Taiwan such as Mrs. Cynthia Lee, Moon Chen, Theodore Way and
several others. We hope they will join us upon our arrival and be
with us for the duration of our conclave.

For your information, I have sent biographies of 40 of the 50-odd
members on the enclosed list to Mr. James Wei as per his request.
The others will follow shortly.

Please advise if there is anything further I can do.

My very best personal regards,

Sincerely,

Leon Spector
President

Encl: List of tour members

1968/07/04
《蔣經國總統文物》 *President Chiang Ching-kuo's Collection*
典藏號 Archive No. ：005-010306-00013-001

圖片：國史館 photo courtesy of Academia Historica

(M) Arther, Wm. C.
(M) Bailey, Addison*
(M) Beggs, Walter
    Beggs, Mrs. Lou
(M) Beneda, Glen
    Beneda, Mrs. Elinor
(M) Bennett, Maj. Gen. Alan
    Bennett, Mrs. Eleanor
(M) Bobnien, Mary
(M) Brown, David J.
    Brown, Mrs. Sharon
(M) Cavanaugh, Cynthia
(M) Cooney, Jack
    Cooney, Mrs. Maybelle
    Cooper, Merrill (TA)
(M) Corkin, Herbert*
    Corkin, Mrs. Ruth
    Corkin, Marjorie
    Corkin, Janice
(M) Chennault, Mrs. Claire L. (T)
(M) Chennault, Col. John S. (T)*
    Chennault, Mrs. Irene (T)
    Chennault, Sharon (T)
    Chennault, Janet (T)
(M) Cox, Paul
    Cox, Mrs. Wilma
(M) Culp, Charles
    Culp, Mrs. Dorothy
    D'Almaine, Thelma
(M) Doyle, Charlton*
    Doyle, Mrs. Dorothy
(M) Duffy, Mrs. Anna
(M) Eisner, Robert
    Eisner, Mrs. Laurel
(M) Fickes, Clyde
    Fickes, Mrs. Helen
(M) Fisher, William O.*
    Fisher, Mrs. Lillian
(M) Floyd, Albert
    Floyd, Mrs. Frances
(M) Gardner, Robert
    Gardner, Mrs. Jean
(M) Gazo, Albert
    Gazo, Mrs. Jessie
(M) Goette, Mrs. Elise
    Goette, Janet
    Goette, Lois
(M) Graves, Col. Frank
    Graves, Mrs. Frank
(M) Greenthal, Charles*
(M) Jenkins, George
(M) Karp, Arthur
    Karp, Mrs. Bette
    Karp, Patrisia

(M) Love, Albert
    Love, Mrs. Nelda
    Mathieu, Bernice (TA)
(M) McQuaid, James
    McQuaid, Mrs. Mae
(M) McShurley, Joseph
    McShurley, Mrs. Mary
    Miller, Mrs. Sylvia
(M) Mooney, Joseph
    Mooney, Mrs. Marjorie
(M) Nagel, Arthur
    Nagel, Mrs. Catherine
    Nagel, Enid
(M) Nowak, Al C.*
    Nowak, Mrs. Mary Ellen
(M) O'Neill, Robert
(M) Owen, Wallace
    Owen, Mrs. Dorothy
(M) Pierson, Jack
    Pierson, Mrs. Ruth
(M) Ponge, Edward*
    Ponge, Mrs. Audrey
    Ponge, Jane
    Ponge, Lorraine
(M) Rees, Carl
    Rees, Mrs. Eleanor
(M) Remer, William
    Remer, Mrs. Bertha
(M) Roberts, Benjamin
    Roberts, Mrs. B. W.
(M) Robertson, Matthew G.
(M) Sigmund, Harold
    Sigmund, Mrs. Dorothy
(M) Silverstien, Dr. Max
    Silverstein, Mrs. Rhoda
    Silverstein, Susan
(M) Simmons, Dr. Gerald
    Simmons, Mrs. Carol
(M) Sklar, William*
    Sklar, Mrs. Pinnie
(M) Smith, Raymond Jr.
    Smith, Mrs. Emma
    Sowry, Mrs.
(M) Spector, Leon*
    Spector, Mrs. Shirley
    Stokes, Mrs. Hope
    Stokes, Ruth (TA)
(M) Stevens, Louise
(M) Stone, Lt. Gen. Charles B. III *
    Stone, Mrs. Ama
    Stone, Claudia
(M) Thomsen, Fred*
    Thomsen, Mrs. Sunbeam
(M) Walters, Robert*

Zhang Qun, the secretary-general to the president, submitted to President Chiang Kai-shek that based on the memorial from Wei Tao-ming, the minister of foreign affair, the 25th annual meeting of the 14th Air Force Association will be held in Taipei. Zhang politely asked Chiang if he and his wife (Madame Chiang) could host a tea party at the Zhongshan Building in Yangmingshan to receive them for their contribution to fighting against the Japanese for the Republic of China during World War II.

張羣呈蔣中正，據魏道明函：第十四航空隊協會第二十五屆年會將於臺北召開，屆時請安排晉謁總統暨夫人。查該隊於二戰期間助我抗日，擬請在陽明山中山樓賜予茶會款待

追懷陳納德將軍 蔣夫人發表紀念文

飛虎將軍陳納德

【本報訊】蔣總統夫人昨天發表對陳納德將軍紀念文。全文如下：

回憶一九四二年二月間，我偕同今總統委員長赴昆明視察飛虎隊。我對這些可愛的青年——以至克多之勇武事蹟，如聖書上「大衛」和「歌利亞」之鬥的那種維護正義的興亙

任務。而今歲月推移，時事變遷，我對此種種記憶猶新。

尤慰欣慰者乃是我曾為他們的榮譽司令並於促成組隊之實現略盡棉薄。而際此撫弄紀念之文仍可想到當時的情景，有些青年是團團他的肥胖的嬰兒臉，有些青年頭戴的頭髮，帶著愉快表情的酒渦，眼神銳利——儀容煥發的犧牲態充分表現一種進取、信心、發憤和勇武莊重的精神。

如此令人敬佩者，主要由於一位入物對他們不可令人欽羨——甚至在抗戰之前，當我出任航空協會秘書長期間，這個人就是我的顧問，時他予我的印象是：具有緘默及高度工作效率能力，過切運用空戰策略和苦心孤詣發揮有限資源之最大效用。在我們

專態急轉直下的發展期中，有若干顧問在實際情況下考驗中有望廉莫及之感，而他確是一位卓越的領導者。能有此選擇，他確未負殷切的期望。

飛虎隊的青年的空軍戰術家，談吐溫和、是一個卓越的，他戀志堅定、待人熱誠。對部屬雖求紀律嚴明，但不緣近苛求。因此，下，他能能做到充滿同氣相投和親如手足的親愛精神。

一九五八年七月，我獲悉他在紐奧爾良歐治納基醫院（Ochsner Foundation Hospital）罹不治之癌症。當時適我訪問美國，應各地講述之邀。我特抽空往探視這位臥病的良友和戰友。我原想於精神的鼓勵和慰藉，但他瀕危之際，所談的並非屬於他「自我憐憫」和「本身」的問題，而是深深關切當前共產蔓延和我們堅毅不拔的精神。這是我們今天紀念和崇敬的人物。

The wife of the president, Song Meiling, published a commemorative message in memory of Flying Tiger General Chennault.

總統夫人宋美齡發表追懷飛虎將軍陳納德的紀念文

1975/03/29
《蔣中正總統文物》*President Chiang Kai-shek's Collection*
典藏號 Archive No.：002-110102-00004-038

圖片：國史館 photo courtesy of Academia Historica

COMMEMORATION OF AMERICA'S BICENTENNIAL
WELCOME FLYING TIGER POST 9957, VFW

Zhao Juyu, minister of the Executive Yuan Veterans
Affairs Council, and the representative of Veterans
of Foreign Combat Association of the United States
Tiger Troops Taiwan Chapter exchange gifts to keep
longevity of the R.O.C.-U.S. friendship (1976).

行政院國軍退除役官兵輔導委員會主任委員
趙聚鈺與美國海外作戰退伍軍人協會飛虎隊
臺灣分會互贈禮物以誌中美友誼長存

1976/07/03
《台灣新生報》 *Taiwan Shin Sheng Daily News*
典藏號 Archive No.：150-031500-0007-035

圖片：國史館 photo courtesy of Academia Historica

August, 1976

Premier Chiang Ching-kuo
Republic of China

Your Excellency;

It was requested that I carve a wood portrait of Chiang Kai-shek
for presentation to you on the occasion of our reunion, 14th Air
Force Association in Taiwan, Republic of China August 2-9, 1976.

This request presented a challenge; I have done many carvings
but not portraits. A few years ago I did do two portraits; one
of Anna Chennault and one of our beloved General, Claire Lee
Chennault, and have not attempted to do a portrait since.
However, I accepted this challenge and by use of a photograph
provided to me by Consulate General, Konsin C. Shah, I carved
this wood portrait especially for you. It is my humble attempt
to create, in wood, the likeness of a great leader of the Chinese
people, Chiang Kai-shek.

I have been told the portrait is a very good likeness of the
photograph I worked from; however, it is you we most wish
to please and it is our hope that this portrait will please
you.

This portrait is presented to you as a token of friendship
by the men and women of the Flying Tigers of the 14th
Air Force Association.

Sincerely,

Albert E. Love, Governor
Board of Governors
14th Air Force Association

Albert E. Love, governor of Board of Governors of the 14th Air Force Association, wrote to Chiang Ching-kuo, premier of the Republic of China, that the Association in Taiwan presented a woodcut portrait of Chiang Kai-shek as a commemoration for Sino-American friendship.

第十四屆空軍協會董事會羅夫函蔣經國：該協會在臺聯合會為紀念彼此友誼，該會飛虎隊男女共同致贈蔣中正木刻像以資紀念

1976/08
《蔣經國總統文物》 *President Chiang Ching-kuo's Collection*
典藏號 Archive No.：005-010502-00441-001

圖片：國史館 photo courtesy of Academia Historica

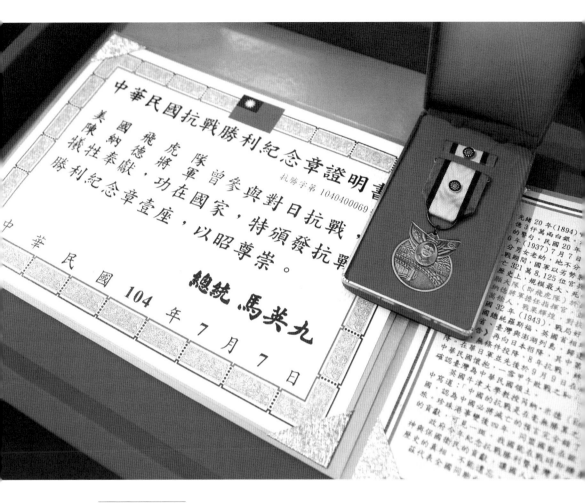

President Ma Ying-jeou received Mrs. Anna Chennault (nee Chen Xiangmei) and issued her the Medal in Commemoration of Victory in the Resistance Against Aggression (2015).

總統馬英九接見美國飛虎隊陳納德將軍遺孀陳香梅女士並頒發抗戰勝利紀念章

2015/10/07

《馬英九總統文物》 *President Ma Ying-jeou's Collection*

典藏號 Archive No.：080-030800-00010-011

圖片：國史館 photo courtesy of Academia Historica

沈呂巡設宴 賀陳香梅 89 歲生日

2014 年 06 月 26 日 04:10 中國時報
劉屏／華盛頓 25 日電

抗戰時期「飛虎航空隊」指揮官陳納德遺孀陳香梅（前右），在華府
雙橡園歡慶 89 歲生日，左為其老伴郝福滿。後排左起：駐美代表沈呂
巡、陳香梅之女陳美麗、沈呂巡夫人李秋萍。遠方立者為前空軍副總
司令陳鴻銓。（駐美代表處提供）

中華民國駐美代表沈呂巡，24 日在雙橡園為抗戰時的「飛虎航空隊」
指揮官陳納德遺孀陳香梅歡慶 89 歲生日。

Reports related to Chen Xiangmei's 89th birthday.
陳香梅 89 歲生日相關報導

2015/10/07
《馬英九總統文物》 *President Ma Ying-jeou's Collection*
典藏號 Archive No.：080-010108-00472-001

圖片：國史館 photo courtesy of Academia Historica

A Brief History of the
Black Cat Squadron

The Black Cat Squadron was a pivotal covert operation that strengthened ROC -U.S. relations during the Cold War. Operating the venerable U2 spy plane across multiple strategic regions, including Mainland China and the Taiwan Strait, the Soviet Union, and North Korea, the squadron, in conjunction with the Central Intelligence Agency (CIA), served as the surveillance arm of the Republic of China Air Force (ROCAF).

Prior to the accession of the U2 spy plane by the Republic of China (ROC), Mao Zedong, the leader of the Chinese Communist Party, had been attempting to consolidate his victory over the Kuomintang (KMT), through military threats and diplomatic isolation. Having fled to the island of Taiwan, the KMT found itself in a precarious political situation. This was compounded by the Communist takeover of the strategic island of Hainan in 1950. By 1951, the territory of the Republic of China had been reduced to just Taiwan and a few outlying islands. The U.S., looking to secure the safety of Taiwan and the stability of the East Asia region, opted to pursue further ties with Chang Kai-shek's government. The culmination of this cooperation resulted in the creation of the Black Cat Squadron.

Also known as the 35th Squadron, the Black Bat Squadron operated out of Taoyuan Air Base in northern Taiwan for thirteen years from 1961 to 1974. Its activities spanned more than 10 million square kilometers, spread out over 30 Chinese provinces. These dangerous missions required its pilots to undergo rigorous training. With just twenty-eight ROCAF pilots trained to operate the U2, with operations often overseen by Chang Kai-shek himself, confidentiality of the project was ensured.

During the 1950s as global tensions were once again on the rise, American strategists in the Pentagon recognized the need for up-to-date intelligence on inland Soviet military installations. Realizing that the only information they had had been confiscated from the Nazis at the end of World War II, plans were quickly drawn up to remedy this deficiency. Those plans eventually culminated in the creation of the U2 Spy Plane.

The intelligence issues surrounding the communist bloc were further compounded by the decision of Chairman Mao Zedong to pursue a Chinese nuclear program. This program would soon become one of the leading reasons for the transfer of the U2 spy plane to the ROCAF. It was hoped that the images gleaned from each overhead flight through mainland China would shed some light into China's secretive weapons programs to forestall future complications that could potentially arise from China's nuclear program.

Nicknamed the Dragon Lady, the Lockheed U2 was designed in 1953, and entered service in 1957. It was an agile, but delicate reconnaissance aircraft: each housing a state-of-the-art Hycon Model B panoramic camera; developed by James Baker. The U2 spy plane was praised for its operational flexibility and aerodynamic design. This was compounded by its adaptable framework and aluminum superstructure: being able to carry up to 3000 pounds. With the U2's average cruising altitude hovering around 70,000 feet, the plane was designed to outrange the most advanced surface-to-air batteries fielded by the U.S.S.R: and while this strategy was successful, advances in Soviet SAM technology eventually made the task much more difficult over time.

In 1958, ROC and American authorities agreed to create the 35th Squadron, which they nicknamed the Black Cats. So sensitive was the nature of these operations, that the joint approval of the U.S. and ROC presidents had to be made before each mission. However, the training of ROC pilots was pushed back due to delays in the transfer of the U2 air frames. Two years later in 1960, 26 ROCAF pilots passed flight training at Laughlin Air Force Base in the United States and subsequently, in January of 1961, the Central Intelligence Agency provided Taiwan with its first two U-2Cs.

Although the usage of the Dragon Lady was not limited to China, with many other flights taking place over North Korea, Vietnam, and even the Soviet Union, the foremost objective of the 35th squadron was to capture photographs of the PRC's nuclear research sites. It was feared by Western military planners that possession of atomic weapons would make an invasion of the Chinese mainland impossible.

The first flight of the Black Cat Squadron's career took place over mainland China in April of 1961. Meanwhile, the 35th's squadron's first year of operation went without incident, but this did not last, as it suffered its first loss on September 9, 1962. Major Chen Huai, traveling on U-2C N.378, was killed in an anti-aircraft missile while flying over Nanchang, China. He was the first of five pilots to die during the Black Cat's thirteen-year operations.

As the Chinese nuclear program grew closer to success, urgency for further intelligence grew with it and after the PRC conducted its third nuclear test on May 9, 1966, operation Tobasco was set into motion by the CIA. Hoping to stay up to date on Chinese nuclear testing, the intention of operation Tobasco was to airdrop a sensor pod into the Taklamakan Desert in China's Xinjiang province.

Though the U2 proved to be a reliable reconnaissance platform and valuable tool to U.S. and ROC intelligence agencies, the aircraft had a short lifespan in Taiwanese service; serving from 1961 to just 1974. The causes of this abrupt service life were due to a multitude of factors. Paramount among these were the advancement of Air Defense capabilities among the Communist Bloc, and the shift in recognition in the United Nations from the Republic of China, to the People's Republic of China by most of the international community.

However, with the new threat of better and more capable Chinese air defenses, together with a gradual political rapprochement between the US and the PRC, the ROC's U2 squadron was severely hindered and was thus prevented from entering Chinese airspace by this new reality. The last U2 flight mission took place over mainland China in 1968. After that point, all missions had to be flown around a buffer zone at least 20 nautical miles around China. Nixon then promised the PRC in 1972 that these flights over Communist China would cease shortly thereafter. Nevertheless, the Black Cat Squadron program continued until May 24, 1974, when Sungchou "Mike" Chiu became the last pilot to fly a 35th Squadron mission over mainland Chinese territory.

These heroic and renowned pilots made history in a span of just thirteen years. From the desolate plains of Manchuria to the tropical forests of Southeast Asia, the Black Cat squadron operated in close conjunction with the CIA during the Cold War. By the end of the ROC's U2 operations, a total of 19 U2 aircraft had been operating in China by the 35th Squadron from 1959 to 1974. In total, the nineteen U2 spy planes that were based at Taoyuan AFB flew 220 sorties altogether: surveying a total of 10 million square kilometers in over 30 provinces in China. However, with five aircraft shot down, three fatalities, and two pilots captured, the success of these reconnaissance missions came at a steep price. Additionally, many more pilots were killed off the Chinese coast or whilst conducting training exercises.

Following the Sino-Soviet split, and President Nixon's subsequent inauguration into the Oval Office, cooperation with ROC was significantly scaled back in favor of reconciliation with Communist China. Following this dramatic shift in America's foreign policy, President Nixon subsequently traveled to Beijing to meet with the aging leader of China at the time, Mao Zedong, and a reset in Sino-American relations was in the offing. Due to political pressure put on the United States by mainland China, President Nixon began rethinking his predecessors' previously harsh approach to dealing with affairs pertaining to mainland China. This led to the Black Cat Squadron to ultimately cease its operations in 1974, following American President Richard Nixon's concerted effort to win over the Chinese Communist leaders and drive a political wedge between the Soviet Union and China. Nixon hoped to exploit this growing rift and benefit American security in the process. Eventually, on June 29, 1974, the remaining U2 aircraft in Taiwanese possession were flown in from the Taoyuan Air Base in the Republic of China to Edwards AFB, California, United States, and turned over to the United States Air Force (USAF).

The legacy of the Black Cat Squadron was an impressive feat of covert action against communist forces, but they reinforced existing bonds and strengthened them even further than before. After retirement, many former pilots of the Black Cat Squadron settled in various locations across the United States, in places such as Las Vegas, Los Angeles, Texas, and Maryland. The crucial military reconnaissance obtained by the Black Cat Squadron, especially the intelligence regarding a Chinese military buildup on the Sino-Soviet border which revealed the stark tensions between the Soviet Union and its communist rival, may have greatly influenced later political deals and maneuverings, such as President Richard Nixon's opening to Communist China and his subsequent visit to Beijing. Almost 50 years later, the CIA has still never fully disclosed its activities in China to the public.

Though the Black Squadron was discontinued almost half a century ago, its legacy continues to be honored today through elaborate ceremonies showcasing the actions of the pilots who partook in the missions. The legacy of these events continues to resonate strongly today.

黑貓中隊

黑貓中隊（The Black Cat Squadron）是冷戰時期強化中華民國與美國軍事合作關係的一項秘密關鍵行動。黑貓中隊是中華民國空軍的高空戰略偵察機部隊，該部隊與美國中央情報局（Central Intelligence Agency）合作，以服役時間極長的 U2 偵察機執行任務，飛越諸多極具戰略意義的地區，包括中國大陸、臺灣海峽、蘇聯、北韓等地進行偵照。

在中華民國參加 U2 偵察機偵照的行動之前，中國共產黨的領導人毛澤東一直試圖以軍事威脅和外交孤立雙管齊下，以求底定中共在國共內戰中的勝利局面。因此中華民國政府在撤往臺灣之後，發覺自身處在十分危殆的境地。一九五〇年，中共拿下具戰略意義的海南島，使得情勢更加嚴峻。到了一九五一年，中華民國的實際管轄範圍只剩下臺灣和一些沿海島嶼。此時，美國出於維護臺灣安全和東亞區域穩定的考慮，選擇與蔣中正政府發展進一步的關係。黑貓中隊的成立，就是雙方合作最密切時期的產物。

黑貓中隊又被稱為第三十五偵察中隊，從一九六一年到一九七四年，該部隊從位於臺灣北部的空軍桃園基地起飛出任務，長達十三年。其活動範圍超過一千萬平方公里，遍布中國三十多個省份。由於這些任務極具危險性，因此飛行員需要接受嚴格的訓練。而接受過 U2 飛行培訓的中華民國空軍飛行員總共只有二十八位，再加上出任務時經常由蔣中正本人親自督導，使得該專案得以保持高度機密。

在一九五〇年代，世界緊張局勢再一次升高，在五角大廈裡制定全球戰略的美國高層這才發覺，他們對於蘇聯內陸軍事設施的情資亟需更新。美方意識到他們手中握有的情資，都還是二次大戰結束時從納粹德國繳獲得來的，因此很快就制定諸多情報計畫以彌補此一缺陷。U2 偵察機便在這些計畫之下應運而生。

美方在獲知毛澤東決心發展核子武器之後，使取得共產國家集團情報的需求更形加劇。中國發展核武計畫很快也成為美國決定將 U2 偵察機移交給中華民國空軍的主要原因。美方希望從每次飛越中國大陸上空的飛行任務中收集到的偵照圖中，能對中共的機密武器發展計畫有所了解，以期得以預先防範將來中國核子計畫造成的複雜紛亂局面。

U2 偵察機的綽號叫「蛟龍夫人」（Dragon Lady），由洛克希德（Lockheed）公司於一九五三年設計開發，在一九五七年投入美國空軍服役。這是一架敏捷卻防禦力脆弱的偵察機：每架 U2 機上都搭載了由詹姆斯・貝克（James Baker）研發、當時最先進的「B」型高空偵照全景攝影相機。U2 偵察機因其操作靈活以及貼合空氣動力學的機身設計而備受讚譽。其適應性強的框架和鋁製的上層結構更能發揮機體的優勢：最高能乘載三千磅的重量。U2 偵察機的平均巡航高度約在七萬英呎（約兩萬一千公尺）左右，這款飛機設計的飛行高度就是要超越蘇聯部署最先進的地對空火炮；儘管這項策略在初期取得成功，但是隨著時間的推移，蘇聯地對空飛彈技術的進步，終究使得 U2 執行任務變得愈來愈困難。

　　一九五八年，中華民國和美方軍事高層同意成立綽號「黑貓中隊」的第三十五偵察中隊。由於黑貓中隊的任務性質極為敏感，因此每次出任務前都要得到中華民國與美國最高領導人的親自批准。然而，由於U2 製造交貨的耽擱，訓練中華民國飛行員的計畫也因而延後。兩年後，也就是一九六〇年，二十六名中華民國空軍飛行員在美國德州的勞夫林空軍基地（Laughlin Air Force Base）通過駕駛 U2 偵察機的飛行訓練，隨後在一九六一年一月，中央情報局交付臺灣首批兩架 U2-C 偵察機。

　　雖然「蛟龍夫人」出任務的地點並不限於中國大陸（該中隊也出勤深入北韓、北越、甚至蘇聯等國上空偵察），不過第三十五偵察中隊最重要的任務，仍然是拍攝中華人民共和國核武研發場所的照片。西方各國制定軍事戰略的高層擔心，一旦中國大陸擁有核武，將使入侵作戰不再可能實現。

　　一九六一年四月，「黑貓中隊」在中國大陸上空進行了成軍以來的首次任務。第三十五中隊第一年出任務過程堪稱平順，沒有發生任何意外，但這樣的情形並未持續太久，該中隊在一九六二年九月九日遭遇了第一次損失。當天，駕駛編號「三七八」U2-C 偵察機的陳懷少校，在江西南昌上空遭到防空飛彈擊落殉職。在「黑貓中隊」十三年任務期間，有五位飛行員陸續在任務中殉職，陳懷是當中的第一位。

隨著中共發展核武計畫愈來愈接近成功，獲取進一步情報的急迫性也日漸升高。在中華人民共和國於一九六六年五月九日進行第三次核彈試爆後，中情局便啟動「菸草行動」。「菸草行動」的目的是將一具感測器夾艙投放於中國新疆省的塔克拉瑪干沙漠，以求掌握中國核武試爆的最新情況。

　　儘管 U2 偵察機對於美國和中華民國的情報機關來說是一可靠的載具，也是極具價值的偵察手段，但是這款飛機在臺灣服役的年限，只由一九六一到一九七四年，並不算長。服役時間之所以戛然而止，背後是多種原因造成的。其中最重要的因素，一是共產主義國家防空技術的提升進步，再者就是聯合國中國代表權問題，大多數國家由原先承認中華民國，轉而承認中華人民共和國。

　　不過，隨著中共防空能力的日見提升，加上美中之間走向政治和解一途的國際新現實，中華民國空軍的 U2 偵察機中隊因而受到嚴重阻礙，無法進入中國大陸領空。一九六八年，U2 偵察機最後一次飛入中國大陸上空執行偵照任務。自此以後，所有任務都必須在中國海岸線至少二十海里外的緩衝區進行。一九七二年，美國總統尼克森（Richard Nixon）訪問中華人民共和國時，更向中方承諾：不久之後將會停止這些對共產中國進行偵查的任務。雖然如此，「黑貓中隊」仍舊持續執行任務，直到一九七四年五月二十四日，由英文名為「麥克」（Mike）的邱松州中校執行了第三十五中隊最後一次對中國大陸的偵照任務。

　　這些英勇而備受尊敬的飛行員，在短短十三年的時間裡創造了歷史。從中國東北荒蕪的平原到東南亞的熱帶森林，「黑貓中隊」在冷戰期間和美國中情局密切合作執行任務。從一九五九年到一九七四年，到了中華民國空軍結束其 U2 偵察任務時，第三十五中隊一共有十九架 U2 在中國執行任務。以桃園軍用機場為基地的十九架 U2 間諜偵察機總共飛行了二百二十架次，飛臨中國大陸三十多個省份，總面積達一千萬平方公里。然而，這些偵察任務的成功也付出了高昂的代價：五架飛機被擊落，三名飛行員殉職，兩人被俘。此外，還有更多飛行員在中國沿海執行任務或進行演習訓練時喪生。

在蘇聯與中共分道揚鑣、尼克森就任美國總統之後，美方大幅縮減與中華民國的合作，並轉向與共產中國和解。美國外交政策發生此一重大轉變之際，尼克森隨即前往北京訪問，會見了當時年事已高的中共領導人毛澤東，中國大陸與美國關係的新階段即將到來。由於中國大陸對美國施加的政治壓力，尼克森開始反思他的前任在處理中國事務上是否太過強硬。尼克森總統極力拉攏中共領導高層，並藉此分化中共與蘇聯之間的關係，這就導致了「黑貓中隊」最終在一九七四年結束任務。尼克森希望利用中蘇之間日益擴大的裂痕，以保障美國的安全。終於，一九七四年六月二十九日，美國方面將剩下的 U2 飛機由空軍桃園基地空運到位於美國加州的愛德華空軍基地（Edwards AFB），並歸建美國空軍。

「黑貓中隊」是昔日針對共產主義勢力應運而生的秘密部隊，它不但完成了令人感佩的任務壯舉，更使得既有的聯繫益發緊密鞏固。「黑貓中隊」的隊員退役之後，散居美國拉斯維加斯、洛杉磯、德州和馬里蘭州等地。「黑貓中隊」當年取得的關鍵情資，特別是關於中國在中蘇邊境集結重兵的情報，揭露出蘇聯與中共之間劍拔弩張的緊張關係，可能對之後的國際政治協定及策略產生了重大的影響，像是理查·尼克森總統向共產中國示好，以及他隨後訪問北京，就是一例。將近五十年後，中情局迄今尚未向公眾完整披露其當年在中國活動的紀錄。

雖然「黑貓中隊」結束任務迄今已經將近半個世紀，但是它對後世帶來的影響，依然在精心設計的紀念活動與陳列展示參與任務飛行員的事蹟中得到彰顯，其遺緒至今仍能引起世人的強烈迴響和共鳴。

Chiang Ching-kuo greeted Soong Meiling (Madame Chiang Kai-shek) who returns from the United States in Taoyuan Air Force Base

蔣經國在空軍桃園基地迎接由美國返國之宋美齡

1950/01/13
《蔣中正總統文物》 *President Chiang Kai-shek's Collection*
典藏號 Archive No.：002-050113-00004-169

圖片：國史館 photo courtesy of Academia Historica

Chiang Kai-shek, The Director-General of the
Chinese Nationalist Party (Kuomintang), went
to the airport to greet Song Meiling (Madame
Chiang) back from the United States in Taoyuan
Air Force Base
中國國民黨總裁蔣中正在空軍桃園基地
親迎宋美齡由美國返國

1950/01/13
《蔣中正總統文物》 *President Chiang Kai-shek's Collection*
典藏號 Archive No.：002-050113-00004-171

圖片：國史館 photo courtesy of Academia Historica

Chiang Kai-shek, The Director-General of the
Chinese Nationalist Party (Kuomintang), went
to the airport to greet Song Meiling (Madame
Chiang) back from the United States in Taoyuan
Air Force Base

中國國民黨總裁蔣中正在空軍桃園基地親
迎宋美齡由美國返國

1950/01/13
《蔣中正總統文物》 *President Chiang Kai-shek's Collection*
典藏號 Archive No.：002-050113-00004-172

圖片：國史館 photo courtesy of Academia Historica

Chiang Wei-kuo greeted Song Meiling who
returned from the United States in Taoyuan Air
Force Base

蔣緯國在空軍桃園基地迎接由美國返國之
宋美齡

1950/01/13
《蔣中正總統文物》*President Chiang Kai-shek's Collection*
典藏號 Archive No.：002-050113-00004-170

圖片：國史館 photo courtesy of Academia Historica

The couples of Chiang Ching-kuo and Chiang
Wei-kuo took a photo while waiting for Song
Meiling (Madame Chiang Kai-shek) at the
airport

蔣經國及蔣緯國兩夫婦於接機等候宋美
齡時留影

1950/01/13
《蔣中正總統文物》 *President Chiang Kai-shek's Collection*
典藏號 Archive No.：002-050113-00029-109

圖片：國史館 photo courtesy of Academia Historica

Chiang Ching-kuo, the chief commissioner of the Chinese Nationalist Party's Taiwan Branch and his wife Faina Chiang Fang-liang took a photo while waiting for Song Meiling (Madame Chiang Kai-shek) at the airport

中國國民黨臺灣省黨部主任委員蔣經國與蔣方良等於接機等候宋美齡時留影

1950/01/13

《蔣經國總統文物》 *President Chiang Ching-kuo's Collection*

典藏號 Archive No.：005-030203-00001-075

圖片：國史館 photo courtesy of Academia Historica

The couples of Chiang Ching-kuo and Chiang Wei-kuo took a photo while waiting for Song Meiling (Madame Chiang Kai-shek) at the airport

蔣經國及蔣緯國兩夫婦於接機等候宋美齡時留影

1950/01/13

《蔣中正總統文物》 *President Chiang Kai-shek's Collection*
典藏號 Archive No.：002-050113-00029-110

圖片：國史館 photo courtesy of Academia Historica

President Chiang Kai-shek visited
Taoyuan Air Force Base
蔣中正總統蒞臨空軍桃園基地

1958/11/27
《蔣中正總統文物》 *President Chiang Kai-shek's Collection*
典藏號 Archive No.：002-050101-00033-184

圖片：國史館 photo courtesy of Academia Historica

President Chiang Kai-shek listens to
Taoyuan Air Force Base briefing
蔣中正總統聽取空軍桃園基地簡報

1958/11/27
《蔣中正總統文物》 *President Chiang Kai-shek's Collection*
典藏號 Archive No.：002-050101-00033-189

圖片：國史館 photo courtesy of Academia Historica

Chiang Ching-kuo, minister of the Veterans
Employment Advisory Committee, accompanied
President Chiang Kai-shek to inspect the
Taoyuan Air Force Base

國軍退除役官兵就業輔導委員會主任委員
蔣經國陪侍蔣中正總統巡視空軍桃園基地

1961/11/09
《蔣經國總統文物》 *President Chiang Ching-kuo's Collection*
典藏號 Archive No.：005-030204-00006-021

圖片：國史館 photo courtesy of Academia Historica

President Chiang Kai-shek inspects
Taoyuan Air Force Base
蔣中正總統巡視空軍桃園基地

1961/11/09
《蔣中正總統文物》 *President Chiang Kai-shek's Collection*
典藏號 Archive No.：002-050101-00044-274

圖片：國史館 photo courtesy of Academia Historica

President Chiang Kai-shek inspects Taoyuan Air Force Base

蔣中正總統巡視空軍桃園基地

1961/11/09
《蔣中正總統文物》 *President Chiang Kai-shek's Collection*
典藏號 Archive No.：002-050101-00044-273

圖片：國史館 photo courtesy of Academia Historica

President Chiang Kai-shek inspects Taoyuan Air Force Base

蔣中正總統巡視空軍桃園基地

1961/11/09
《蔣中正總統文物》 *President Chiang Kai-shek's Collection*
典藏號 Archive No.：002-050101-00044-272

圖片：國史館 photo courtesy of Academia Historica

President Chiang Kai-shek inspects
Taoyuan Air Force Base
蔣中正總統巡視空軍桃園基地

1961/11/09
《蔣中正總統文物》 *President Chiang Kai-shek's Collection*
典藏號 Archive No.：002-050101-00044-271

圖片：國史館 photo courtesy of Academia Historica

President Chiang Kai-shek inspects Taoyuan Air Force Base

蔣中正總統巡視空軍桃園基地

1961/11/09
《蔣中正總統文物》 *President Chiang Kai-shek's Collection*
典藏號 Archive No.：002-050101-00044-275

圖片：國史館 photo courtesy of Academia Historica

President Chiang Kai-shek inspects Taoyuan
Air Force Base
蔣中正總統巡視空軍桃園基地

1961/11/09
《蔣中正總統文物》 *President Chiang Kai-shek's Collection*
典藏號 Archive No.：002-050101-00044-276

圖片：國史館 photo courtesy of Academia Historica

President Chiang Kai-shek visited
Taoyuan Air Force Base aerial photograph
interpretation operations

蔣中正總統參觀空軍桃園基地空照判
讀作業

1961/11/09
《蔣中正總統文物》 *President Chiang Kai-shek's Collection*
典藏號 Archive No.：002-050101-00044-283

圖片：國史館 photo courtesy of Academia Historica

President Chiang Kai-shek visited Taoyuan Air Force Base aerial photograph interpretation operations

蔣中正總統參觀空軍桃園基地空照判讀作業

1961/11/09
《蔣中正總統文物》 *President Chiang Kai-shek's Collection*
典藏號 Archive No.：002-050101-00044-284

圖片：國史館 photo courtesy of Academia Historica

Chiang Ching-kuo, minister of the Veterans
Employment Advisory Committee,
accompanied President Chiang Kai-shek to
inspect the Taoyuan Air Force Base

國軍退除役官兵就業輔導委員會主任
委員蔣經國陪侍蔣中正總統巡視空軍
桃園基地

1961/11/09
《蔣經國總統文物》*President Chiang Ching-kuo's Collection*
典藏號 Archive No.：005-030204-00006-021

圖片：國史館 photo courtesy of Academia Historica

President Chiang Kai-shek visited
Taoyuan Air Force Base aerial photograph
interpretation operations
蔣中正總統參觀空軍桃園基地空照判
讀作業

1961/11/09
《蔣中正總統文物》 *President Chiang Kai-shek's Collection*
典藏號 Archive No.：002-050101-00044-285

圖片：國史館 photo courtesy of Academia Historica

Accompanied by Chiang Ching-kuo, minister
of Veterans Employment Advisory Committee ,
Vice-President Chen Cheng went to Taoyuan Air
Force Base to inspect the Special Task Units

陳誠副總統由國軍退除役官兵就業輔導委
員會主任委員蔣經國陪同赴空軍桃園基地
視察空軍特種單位（照相技術隊）

1962/08/03
《陳誠副總統文物》 *Vice-President Chen Cheng's Collection*
典藏號 Archive No.：008-030604-00027-001

圖片：國史館 photo courtesy of Academia Historica

Accompanied by Chiang Ching-kuo, minister of Veterans Employment Advisory Committee , Vice-President Chen Cheng went to Taoyuan Air Force Base to inspect the Special Task Units

陳誠副總統由國軍退除役官兵就業輔導委員會主任委員蔣經國陪同赴空軍桃園基地視察空軍特種單位（照相技術隊）

1962/08/03
《陳誠副總統文物》 *Vice-President Chen Cheng's Collection*
典藏號 Archive No.：008-030604-00027-001

圖片：國史館 photo courtesy of Academia Historica

**Accompanied by Chiang Ching-kuo, minister of
Veterans Employment Advisory Committee , Vice-
President Chen Cheng went to Taoyuan Air Force
Base to inspect the Special Task Units**

陳誠副總統由國軍退除役官兵就業輔導委
員會主任委員蔣經國陪同赴空軍桃園基地
視察空軍特種單位（照相技術隊）

1962/08/03
《陳誠副總統文物》 *Vice-President Chen Cheng's Collection*
典藏號 Archive No.：008-030604-00027-001

圖片：國史館 photo courtesy of Academia Historica

**Accompanied by Chiang Ching-kuo, minister of
Veterans Employment Advisory Committee , Vice-
President Chen Cheng went to Taoyuan Air Force
Base to inspect the Special Task Units**

陳誠副總統由國軍退除役官兵就業輔導委
員會主任委員蔣經國陪同赴空軍桃園基地
視察空軍特種單位（照相技術隊）

1962/08/03
《陳誠副總統文物》 *Vice-President Chen Cheng's Collection*
典藏號 Archive No.：008-030604-00027-001

圖片：國史館 photo courtesy of Academia Historica

Accompanied by Chiang Ching-kuo, minister
of Veterans Employment Advisory Committee ,
Vice-President Chen Cheng went to Taoyuan Air
Force Base to inspect the Special Task Units

陳誠副總統由國軍退除役官兵就業輔導委
員會主任委員蔣經國陪同赴空軍桃園基地
視察空軍特種單位（照相技術隊）

1962/08/03
《陳誠副總統文物》 *Vice-President Chen Cheng's Collection*
典藏號 Archive No.：008-030604-00027-001

圖片：國史館 photo courtesy of Academia Historica

Accompanied by Chiang Ching-kuo, minister of Veterans Employment Advisory Committee, Vice-President Chen Cheng went to Taoyuan Air Force Base to inspect the Special Task Units

陳誠副總統由國軍退除役官兵就業輔導委員會主任委員蔣經國陪同赴空軍桃園基地視察空軍特種單位（照相技術隊）

1962/08/03
《陳誠副總統文物》 *Vice-President Chen Cheng's Collection*
典藏號 Archive No.：008-030604-00027-001

圖片：國史館 photo courtesy of Academia Historica

Accompanied by Chiang Ching-kuo, minister of Veterans Employment Advisory Committee, Vice-President Chen Cheng went to Taoyuan Air Force Base to inspect the Special Task Units

陳誠副總統由國軍退除役官兵就業輔導委員會主任委員蔣經國陪同赴空軍桃園基地視察空軍特種單位（照相技術隊）

1962/08/03
《陳誠副總統文物》 *Vice-President Chen Cheng's Collection*
典藏號 Archive No.：008-030604-00027-001

圖片：國史館 photo courtesy of Academia Historica

Accompanied by Chiang Ching-kuo, minister
of Veterans Employment Advisory Committee ,
Vice-President Chen Cheng went to Taoyuan Air
Force Base to inspect the Special Task Units

陳誠副總統由國軍退除役官兵就業輔導委
員會主任委員蔣經國陪同赴空軍桃園基地
視察空軍特種單位（照相技術隊）

1962/08/03
《陳誠副總統文物》 *Vice-President Chen Cheng's Collection*
典藏號 Archive No.：008-030604-00027-001

圖片：國史館 photo courtesy of Academia Historica

Accompanied by Chiang Ching-kuo, minister
of Veterans Employment Advisory Committee ,
Vice-President Chen Cheng went to Taoyuan Air
Force Base to inspect the Special Task Units

陳誠副總統由國軍退除役官兵就業輔導委
員會主任委員蔣經國陪同赴空軍桃園基地
視察空軍特種單位（照相技術隊）

1962/08/03
《陳誠副總統文物》 *Vice-President Chen Cheng's Collection*
典藏號 Archive No.：008-030604-00027-001

圖片：國史館 photo courtesy of Academia Historica

**Accompanied by Chiang Ching-kuo, minister
of Veterans Employment Advisory Committee ,
Vice-President Chen Cheng went to Taoyuan Air
Force Base to inspect the Special Task Units**

陳誠副總統由國軍退除役官兵就業輔導委
員會主任委員蔣經國陪同赴空軍桃園基地
視察空軍特種單位（照相技術隊）

1962/08/03
《陳誠副總統文物》 *Vice-President Chen Cheng's Collection*
典藏號 Archive No.：008-030604-00027-001

圖片：國史館 photo courtesy of Academia Historica

Accompanied by Chiang Ching-kuo, minister
of Veterans Employment Advisory Committee ,
Vice-President Chen Cheng went to Taoyuan Air
Force Base to inspect the Special Task Units

陳誠副總統由國軍退除役官兵就業輔導委
員會主任委員蔣經國陪同赴空軍桃園基地
視察空軍特種單位（照相技術隊）

1962/08/03
《陳誠副總統文物》 *Vice-President Chen Cheng's Collection*
典藏號 Archive No.：008-030604-00027-001

圖片：國史館 photo courtesy of Academia Historica

Accompanied by Chiang Ching-kuo, minister
of Veterans Employment Advisory Committee ,
Vice-President Chen Cheng went to Taoyuan Air
Force Base to inspect the Special Task Units

陳誠副總統由國軍退除役官兵就業輔導委
員會主任委員蔣經國陪同赴空軍桃園基地
視察空軍特種單位（照相技術隊）

1962/08/03
《陳誠副總統文物》 *Vice-President Chen Cheng's Collection*
典藏號 Archive No.：008-030604-00027-001

圖片：國史館 photo courtesy of Academia Historica

Accompanied by Chiang Ching-kuo, minister
of Veterans Employment Advisory Committee ,
Vice-President Chen Cheng went to Taoyuan Air
Force Base to inspect the Special Task Units

陳誠副總統由國軍退除役官兵就業輔導委
員會主任委員蔣經國陪同赴空軍桃園基地
視察空軍特種單位（照相技術隊）

1962/08/03
《陳誠副總統文物》 *Vice-President Chen Cheng's Collection*
典藏號 Archive No.：008-030604-00027-001

圖片：國史館 photo courtesy of Academia Historica

President Chiang Kai-shek and Madame
Chiang inspect Taoyuan Air Force Base, having
picture with Li Zhi, the wing commander and
other personnel.
蔣中正總統伉儷於空軍桃園基地與聯隊
長李碩等人合影

1963/12/18
《蔣中正總統文物》 *President Chiang Kai-shek's Collection*
典藏號 Archive No.：002-050101-00053-211

圖片：國史館 photo courtesy of Academia Historica

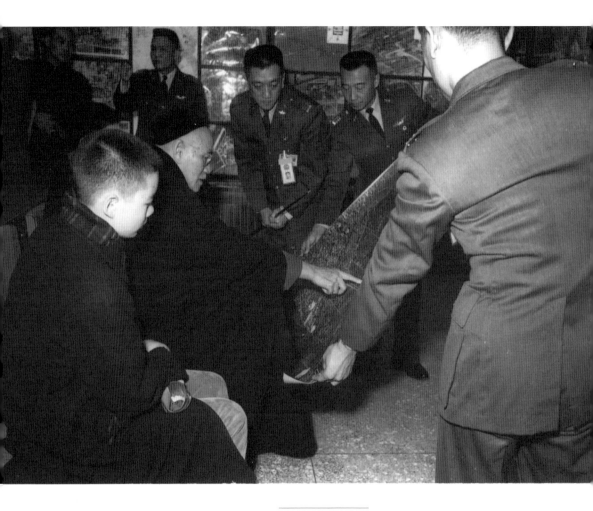

President Chiang Kai-shek and Madame
Chiang inspect Taoyuan Air Force Base
蔣中正總統巡視空軍桃園基地

1963/12/18
《蔣中正總統文物》 *President Chiang Kai-shek's Collection*
典藏號 Archive No.：002-050101-00053-209

圖片：國史館 photo courtesy of Academia Historica

President Chiang Kai-shek and Madame Chiang inspect Taoyuan Air Force Base

蔣中正總統巡視空軍桃園基地

1963/12/18
《蔣中正總統文物》 *President Chiang Kai-shek's Collection*
典藏號 Archive No.：002-050101-00053-208

圖片：國史館 photo courtesy of Academia Historica

President Chiang Kai-shek and Madame
Chiang inspect Taoyuan Air Force Base
蔣中正總統巡視空軍桃園基地

1963/12/18
《蔣中正總統文物》*President Chiang Kai-shek's Collection*
典藏號 Archive No.：002-050101-00053-210

圖片：國史館 photo courtesy of Academia Historica

President Chiang Kai-shek and Madame
Chiang inspect Taoyuan Air Force Base

蔣中正總統巡視空軍桃園基地

1963/12/18
《蔣中正總統文物》 *President Chiang Kai-shek's Collection*
典藏號 Archive No.：002-050101-00108-046

圖片：國史館 photo courtesy of Academia Historica

Taiwan Provincial Governor Hsieh Tung-min led the magistrates and mayors of the province to visit the Taoyuan Air Force Base

臺灣省政府主席謝東閔率領全省各縣市長訪問空軍桃園基地

1975/01/07
《謝東閔副總統文物》 *Vice-President Hsieh Tung-min's Collection*
典藏號 Archive No.：009-030204-00050-011

圖片：國史館 photo courtesy of Academia Historica

Taiwan Provincial Governor Hsieh Tung-min
led the magistrates and mayors of the province
to visit the Taoyuan Air Force Base

臺灣省政府主席謝東閔率領全省各縣
市長訪問空軍桃園基地

1975/01/07
《謝東閔副總統文物》 *Vice-President Hsieh Tung-min's Collection*
典藏號 Archive No.：009-030204-00050-011

圖片：國史館 photo courtesy of Academia Historica

Cai Shengxiong, pilot of the 35th squadron of the Republic of China Air Force (Black Cat squadron), took photo next to a U-2R tactical reconnaissance aircraft.

中華民國空軍第三十五偵查中隊（黑貓中隊）隊員蔡盛雄與 U-2R 型戰略偵察機合影

1973

圖片：喆閎人文工作室
Photo courtesy of Zhehong Humanities Studio

参考資料
Bibliography

專著 Published Work

1. Jackson, Daniel. Fallen Tigers: The Fate of America's Missing Airmen in China during World War II. Lexington: KT, University Press of Kentucky, 2020.

2. Kleiner, Sam. The Flying Tigers: The Untold Story of the American Pilots Who Waged a Secret War Against Japan. New York: Penguin Random House, 2018

3. Pedlow, Gregory, and Donald Welzenbach. The Central Intelligence Agency and Overhead Reconnaissance: The U-2 and OXCART Programs, 1954-1974. New York: Skyhorse Publishing, 2016.

網路資料 Website Data

1. "1972 visit by Richard Nixon to China," Wikipedia, https://reurl.cc/5O1Qoy, 8/25/2023

2. "Central Intelligence Agency," Wikipedia, 10/3/2023

3. "Chiang Kai-shek," Wikipedia, https://reurl.cc/dmDgXz, 9/28/2023

4. "Incident Lockheed U-2C 358," Aviation Safety Network, https://reurl.cc/Y0epLx, 9/7/2023

5. "The Black Cat Squadron - Cold War spy flights over mainland China from Taiwan," 2 November 2022, School of Oriental and African Studies, University of London. https://reurl.cc/RyzmVe, 9/8/2023

6. "The Black Cat Squadron," 9 March 2023, Air Force of Republic of China. https://reurl.cc/l7DArA, 9/10/2023

7. "U-2S/TU-2S," Air Force, https://reurl.cc/2ELe4a, 9/1/2023

8. A Brief History of The US Navy PBY Black Cats Squadron (WWII), Cockpit USA, https://reurl.cc/A0Aa7j, 9/30/2023

9. Alan Taylor, "Remembering the Korean War," The Atlantic, July 27, 2016.

10. Black Cat Squadron, Wikipedia, https://reurl.cc/RyzgrZ, 7/28/2023

11. Bob Bergin, "The Growth of China's Air Defenses: Responding to Covert Overflights, 1949–1974 - Central Intelligence Agency," https://web.archive.org/web/20140104204338/https://www.cia.gov/library/center-for-the-study-of-intelligence/csi-publications/csi-studies/studies/vol-57-no-2/the-growth-of-china2019s-air-defenses-responding-to-covert-overflights-194920131974.html., 9/6/2023

12. File: Taiwan location map.svg, Wikipedia, https://reurl.cc/a4VgNY, 9/30/2023

13. FLYING THE HUMP DURING WORLD WAR II, Lyon Air Museum, https://reurl.cc/kaX2Ed, 7/28/2023

14. Jeff Neal, "When Nixon went to China," Feb 17, 2022, Harvard Law Today, https://hls.harvard.edu/today/when-nixon-went-to-china/ 8/30/2023

15. Li Wang, The Museum of Flight, Cold War Reconnaissance Squadrons. YouTube, 1 Mar. 2017, https://reurl.cc/7MkNDy, 7/30/2023.

16. Nixon's China Visit, 1972, PBS, https://reurl.cc/E1oy8a 10/3/2023

17. Taklamakan Desert Facts & Information - Beautiful World Travel Guide, Facts & Information - Beautiful World Travel Guide, https://reurl.cc/XmEgWD, 8/20/2023

18. The End of an Era, TaiwanAirBlog, https://reurl.cc/a4VgrG ,9/7/2023

19. U-2 Operations: Aircraft Assigned, TaiwanAirPower, https://web.archive.org/, 9/8/2023

20. U-2 Operations: Pilots, TaiwanAirPower, https://web.archive.org/, 9/8/2023

21. WE.177, Wikipedia, https://reurl.cc/OjvgkR, 9/29/2023

22. When China shot down five U-2 spy planes at the height of the Cold War, CNN, https://reurl.cc/E1oyrv, 9/17/2023

國家圖書館出版品預行編目 (CIP) 資料

翱翔天際：圖說飛虎隊與黑貓中隊 = Sky high : a
photographic history of the Flying Tigers and the
Black Cat Squadron / Bennett Quo(郭代忠), Everett
Wang(王艾唯), Nathan Yao, Patrick Hao(郝健坤) 著 .
-- 初版 . -- 新北市 : 詰閎人文工作室 , 2023.10
　　面 ;　　公分 . -- (歷史影像 ; 1)
ISBN 978-986-99268-4-3(平裝)

1.CST: 空軍 2.CST: 中華民國軍事 3.CST: 軍事史

598.8   112017306

歷史影像 1

翱翔天際：圖說飛虎隊與黑貓中隊

Sky High: A Photographic History of the Flying Tigers and the Black
Cat Squadron

主編 Edit / 楊善堯 Yang, Shan-Yao
作者 Author / Bennett Quo(郭代忠)、Everett Wang(王艾唯)、Nathan Yao、Patrick Hao(郝健坤)
翻譯校對 Translation and proofreading / 廖彥博 Liao, Yen-Po
設計排版 Design Layout / 吳姿穎 Wu, Tzu-Ying

出版 Publish / 詰閎人文 ZHEHONG HUMANITIES STUDIO
地址 Address / 242011 新北市新莊區中華路一段 100 號 10 樓
　　　　　　　 10F., No. 100, Sec. 1, Zhonghua Rd., Xinzhuang Dist., New Taipei City 242011 ,
　　　　　　　 Taiwan (R.O.C.)
電話 Telephone / +886-2-2277-0675
信箱 Email / zhehong100101@gmail.com
網站 Website / http://zhehong.tw/

初版一刷 First Edition Brush / 2023 年 10 月
定價 Pricing / 新臺幣 NT$ 300 元、美元 USD$ 10 元
ISBN / 978-986-99268-4-3
印刷 Print / 秀威資訊科技股份有限公司 Showwe Taiwan